The Legacy of La Luz

La Luz plaza
(Courtesy of Jim See)

The Legacy of La Luz

Building Community with Respect for the Land

Anne Taylor, PhD
and
Cynthia Lewiecki-Wilson, PhD

SUNSTONE PRESS
SANTA FE

© 2018 by Anne Taylor and Cynthia Lewiecki-Wilson
All Rights Reserved
No part of this book may be reproduced in any form or by any electronic or mechanical means including information storage and retrieval systems without permission in writing from the publisher, except by a reviewer who may quote brief passages in a review.

Sunstone books may be purchased for educational, business, or sales promotional use. For information please write: Special Markets Department, Sunstone Press, P.O. Box 2321, Santa Fe, New Mexico 87504-2321.

Cover photograph by Jerry Goffe, Nature Photoworks
Body typeface › Bodoni
Printed on acid-free paper
∞

Library of Congress Cataloging-in-Publication Data

Names: Taylor, Anne, 1933- author. | Lewiecki-Wilson, Cynthia, author.
Title: The legacy of La Luz : building community with respect for the land / by Anne Taylor and Cynthia Lewiecki-Wilson.
Description: Santa Fe : Sunstone Press, 2018. | Includes bibliographical references and index.
Identifiers: LCCN 2018037306 | ISBN 9781632932433 (softcover : alk. paper)
Subjects: LCSH: La Luz (Albuquerque, N.M.)--History. | Predock, Antoine.--Criticism and interpretation. | Architecture--Arid regions--New Mexico--Albuquerque--History--20th century. | Architecture--Environmental aspects--New Mexico--Albuquerque--History--20th century. | Housing development--New Mexico--Albuquerque--History--20th century. | Albuquerque (N.M.)--Buildings, structures, etc.
Classification: LCC NA7117.A74 T39 2018 | DDC 720.9789/610904--dc23
LC record available at https://lccn.loc.gov/2018037306

WWW.SUNSTONEPRESS.COM
SUNSTONE PRESS / POST OFFICE BOX 2321 / SANTA FE, NM 87504-2321 /USA
(505) 988-4418 / ORDERS ONLY (800) 243-5644 / FAX (505) 988-1025

Dedication

We dedicate this book to the developer of La Luz del Oeste, Ray A. Graham, III, and to its architect, Antoine Predock.

Ray Graham and Antoine Predock 2018
(Courtesy of James C. Wilson)

Contents

List of Illustrations / 8

Acknowledgements / 10

1
"A Basic Attitude toward the Land" / 16

2
Adobe Mud and River Trees, A Part
of the Environment, Not Apart from It / 23

3
Community, An Experiment in Democracy / 40

4
Sustaining the Light on the Mesa / 61

Afterword:
Ecosophy, A New Philosophy Applied to La Luz / 73

Notes / 75

Bibliography / 77

Index / 81

About the Authors / 83

List of Illustrations

1. La Luz plaza (Courtesy of Jim See)
2. Ray Graham and Antoine Predock 2018 (Courtesy of James C. Wilson)
3. La Luz pool looking eastward (Courtesy of Jonathan Abdalla)
4. Chaco Canyon's Pueblo Bonito (Courtesy of James C. Wilson)
5. Cutting into the mesa (Courtesy of Ray Graham)
6. The University of New Mexico library (Anne Taylor, author's collection)
7. John Gaw Meem's entryway (Anne Taylor, author's collection)
8. Taos pueblo (Courtesy of James C. Wilson)
9. La Luz (Courtesy of Jim See)
10. Transplanting trees from the bosque (Courtesy of Ray Graham)
11. Units flow with the land (Courtesy of Robert Peters)
12. Pouring concrete (Courtesy of Ray Graham)
13. Building with adobe (Courtesy of Ray Graham)
14. Cluster of units, perimeter roads (Courtesy of Ray Graham)
15. A plaza fountain (Cynthia Lewiecki-Wilson, author's collection)
16. A patio (Courtesy of Robert Peters)
17. Constructing with adobe (Courtesy of Ray Graham)
18. La Luz site map (Courtesy of La Luz Landowners Association)
19. Architect's drawing (Courtesy of Antoine Predock)
20. Interior view (Courtesy of Jim See)
21. A unit on Berm Street (Courtesy of James C. Wilson)
22. Plaza fountain looking toward mountains (Courtesy of James C. Wilson)
23. Pueblo Bonito, Chaco Canyon (Courtesy of James C. Wilson)
24. Memorial Day picnic 2018 (Courtesy of James C. Wilson)
25. Memorial Day picnic (Courtesy of Helen Reilly)
26. Fourth of July picnic (Courtesy of Jennifer Fenstermacher)
27. Budagher Hall, Bosque School 2018 (Courtesy of Elliot Madriss)
28. Quad from Budagher Hall (Courtesy of Elliot Madriss)
29. Tract map (Courtesy of La Luz Landowners Association)
30. Jazz concert on a plaza (Cynthia Lewiecki-Wilson, author's collection)
31. Double rainbow over La Luz (Courtesy of James C. Wilson)
32. Storm on La Luz skyline (Courtesy of Helen Reilly)
33. Clouds and balloon over La Luz (Courtesy of Helen Reilly)
34. A summer day (Anne Taylor, author's collection)

35. Resident with her dog (Courtesy of Jonathan Abdalla)
36. Picuris Pueblo Dancer (Courtesy of James C. Wilson)
37. Picuris Pueblo blessing (Courtesy of Carol Bennett)
38. Community blessing dance, Memorial Day (Courtesy of James C. Wilson)
39. Traditional bluegrass landscape (Courtesy of Jim See)
40. New landscape of native grasses and plants (Courtesy of James C. Wilson)
41. Landscape Master Plan (Courtesy of La Luz Landowners Association)
42. La Luz vegetable garden (Courtesy of James C. Wilson)
43. Coyote beside wall of La Luz (Courtesy of Carol Bennett)
44. Water from caneleta (Courtesy of James C. Wilson)
45. Bosque School students studying the riparian ecosystem 2018 (Courtesy of Elliot Madriss)
46. BEMP 7th grade science (Courtesy of Elliot Madriss)
47. A child climbing zome (Courtesy of Anne Fitzpatrick)
48. A child hanging from zome (Cynthia Lewiecki-Wilson, author's collection)
49. Zonohedron close up (Cynthia Lewiecki-Wilson, author's collection)
50. Graphics (Courtesy of Meredith Taylor)
51. Authors, Anne Taylor and Cynthia Lewiecki-Wilson (Courtesy of Jennifer Fenstermacher)

Acknowledgements

Ray Graham, a young twenty-eight year old in 1967, had the dream of building La Luz. Graham, along with architect Antoine Predock and Didier Raven, were excited about new approaches to land development and innovative ways to design, build and live with the land in clustered communities. Fifty years later, Graham's dream is still alive, though maybe not as extensive as he originally planned. We wish to thank Ray for giving us his history of La Luz over two separate interviews in the spring of 2018. Ray was a pioneer, supporting an architect and small crew of devoted workers to accomplish this dream. Now in his late seventies he can look back with pride, as we do in this book, for giving the world a prototypical development, a gift of community that is a unique place, radiating a spirit of joy and light.

Antoine Predock, the man who designed La Luz, has in the intervening fifty years become an internationally renowned architect. He has received the American Institute of Architects Gold Medal and the Smithsonian Cooper-Hewitt Lifetime Achievement Award. On a sunny April morning, the authors felt fortunate to have a warm heartfelt interview with him. Surrounded by an inventory of hundreds of his models, quick sketches, watercolors, smudged pastels and more, we felt very privileged to be with this talented and productive man. His legacy of creative work is housed in the space where he once lived for many years. Next door are more giant models in his former production studio, now the Antoine Predock Center for Design and Research, part of the University of New Mexico's School of Architecture and Planning. We felt awed as we experienced his joie de vivre and listened as he recounted his history of what he calls the first of his many "landscapers," a term for buildings that embrace the land and are not vertically piercing the sky as skyscrapers.

We were surprised when Predock spontaneously instructed Siri on his cell phone to call France, where Didier Raven now lives. Antoine made instant contact and introduced us to Didier, who over the years has remained a good friend and colleague. And so our interview with Didier Raven was arranged.

On a lovely May day we used Face Time to talk with Didier, now a retired business owner and restaurateur living in the south of France. He described how he and Antoine Predock met and became life-long friends. He recalled the dinner in Corrales when Didier and Ray first talked about a new kind of housing, contemporary and innovative, made of local materials, a mixed-use urban development intended for moderate-income families. Raven described his part in the construction of La Luz and the early years living side-by-side with Graham and Predock, each with their families and their children playing together. We thank Didier for his generosity in talking with us for a full hour across many time zones.

Other invaluable links to the history and community of La Luz have enabled this book. First and foremost, we must thank Jonathan Abdalla, La Luz Office Manager, for generously sharing his photographs and memory of past events. We thank Patrick Gallagher for arranging an interview with inventor Steve Baer and Steve for talking with us. We also thank the many La Luz residents whom we interviewed: Beth Baurick, Marianne Barlow, Arlo and Jette Braun, Brenda Broussard, Hank Botts, Laura Campbell, Anne and Tom Fitzpatrick, Patrick Gallagher, Kathryn Kaminski, Joanne Kimmey, Betsy King, Jimmie Leuder, Sandy Masson, Helen Marsee, Betsy King, Lynn Perls, Robert Peters. We especially thank Marilyn O'Leary, Barbara Thiele, James C. Wilson, and Robert Peters who read drafts of this book and helped to improve it.

Finally, and importantly, we wish to thank the Board of Directors of the La Luz Landowners Association for providing funds in support of this book, and Joan Mirabal, who organized and collected donations, along with Brian and Barbara Thiele. We especially thank the following donors: Forrest and Kathy Adams, an anonymous donor, John and Heather Badal, Marianne and Richard Barlow, Richard and Beth Baurick, Hank Botts, Tim and Jackie Bowen, Arlo and Jette Braun, David and Peggy Breault, Peter and Linda Collins, Nancy Conrow, John and Lynda Elliott, Mark Ennen, Tom and Anne Fitzpatrick, Deborah Kendall-Gallagher and Patrick Gallagher, Randy Gleason, Colonel and Mrs. Ronan C. Grady, Jr., Virginia Hanratty, Matt and Ginny Hautau, Richard and Sheila Hills, Dan Jensen and Terry Williams, Kathryn Kaminsky and Tom Singleton, Jim and Marjorie Kannolt, Betsy King, Matt King, Realtor Chris Lucas, Jimmie Lueder, Glenn Mallory, David and Helen Marsee, Jean Martin, Al and Joan Mirabal, Jane Moody, Joseph and Mary Mucci, Alan and Michele Newport, Marilyn O'Leary and Helen Reilly, Judith Pacht, Calla Ann Pepmueller, Robert Peters, Walter Putnam and Yoko Zeigler, Anthony and Cintra Strippoli, Anne Taylor, Brian and Barbara Thiele, James C. Wilson and Cynthia Lewiecki-Wilson.

La Luz pool looking eastward
(Courtesy of Jonathan Abdalla)

1

"A Basic Attitude toward the Land"[1]

A swimmer doing the backstroke in La Luz's pool can savor New Mexico's intense blue sky, wave to the multicolored balloons hovering over the nearby bosque, or count the many jet trails etching that sky, their white streams leaving 33,000 foot ribbons as they cross our state, probably on their way to more populated urban areas.

If that plane is descending toward Albuquerque, those on board might be struck by the barrenness of the high and low desert, the large open spaces, muddy rivers cutting through irrigation ditches to green farms, flat topped mesas, mountains that rise through several life zones with diverse colored strata and textural patterns. Visitors might also see some of the still existing nineteen pueblos in New Mexico, now also dotted with more contemporary separate single family housing. There are still many square miles, filled with ruins of ancient buried pueblos, not yet, or maybe never to be, excavated.

Our travelers will sense that this land is truly a magical and powerful place. It has been inhabited for thousands of years, and like Chaco Canyon at its height (850-1150 CE), some current architecture still carries the imprint of this strong regionalism, unique in America.

New Mexico has a rich history of sparse and minimal architecture. Building materials of stone, mud and straw come from mother earth, fashioned by hand into building blocks of stone, or adobes, or cut from the earth into terrones, large blocks of earth with embedded grass. Such materials have been used for thousands of years, to make walls for rectilinear or round houses and kivas that are topped with flat roofs. Ramadas and organically designed structures shade the strong mid-day sun. Though sparse appearing, New Mexico's desert geography has been transformed by humans into shelter that embraces and nurtures the human body as well as spirit.

This was the architectural style rooted in the earth that housed Ancestral Puebloan people for thousands of years. Stone and adobe are still the preferred building material in contemporary pueblos, though many houses are now constructed of wood and finished with a stucco exterior. More than a style, this was a way of building that honored the origination stories of their ancestors. Because of its minimal form that rises out of the earth, it complemented well the surrounding natural and geographic elements including the color, adobe brown.

New Mexico has always been a land of travelers and migrants. Ancestral puebloans, before the arrival of the Spanish, built stone and adobe communities and then when water became scarcer or warfare more intense they moved, often establishing new settlements on top of older ones.

When the Spanish arrived in the sixteenth and seventeenth centuries, they settled in small, dispersed farming communities from the provincial capital, La Villa Real de la Santa Fe de San Francisco de Asís, southward along the Rio Grande. In response to brutality and warfare against the native population, puebloans drove the Spanish out of the "province" to El Paso in 1680. When the Spanish reestablished their control twelve years later, they recognized the pueblos as permanent cities and began extending them land rights.

Chaco Canyon's Pueblo Bonito
(Courtesy of James C. Wilson)

In the nineteenth and twentieth centuries, newcomers arrived in New Mexico from the East and Midwest, first via wagon trains and then in the late 1870s by railroad. They came to pursue new business opportunities, such as entrepreneur Fred Harvey who established a chain of hotels; others sought the sunny climate to treat tuberculosis. Modernist artists and social reformers (many of them wealthy women) of the 1920s and 30s came to create new kinds of communities. Migrant farmers of the dust bowl followed the "mother road," Route 66, that took them through Albuquerque, and later the new interstate highway system brought the expanding middle classes in the post World War II period. Many of these newcomers preferred buildings that resembled Eastern seaboard or Eurocentric styles of homes and offices, and they, in turn, added their flavor to New Mexico's indigenous and colonial Spanish architecture.

La Luz del Oeste, a Planned Area Development

It is this multi-cultural and geographic context that sets the stage for one of the most innovative designs and housing projects of our time, beginning in 1968. This project is called La Luz (The Light) del Oeste.

The planned community of La Luz began with the first construction of adobe townhouses on Albuquerque's west mesa in 1968—with units added in 1970, 1974, and 1976. Ray A. Graham III, President of Ovenwest Corporation, bought the land and hired architect Antoine Predock, contractor Gunnar Dahlquist, and later architect Hildreth Barker, to create this forward-looking, environmentally friendly project honoring privacy, shared communal space, and the vast New Mexican landscape of mesas and mountains and rivers.

In 1968 New Mexico was ripe for experimentation. It was a sparsely populated state with its major city, Albuquerque, less than half its current size. Drawn to the state's beauty, quality of light, openness, and diverse cultures, young people from all over the United States came to roam its back roads and artist colonies such as Taos and Santa Fe. Communes dotted the landscape and all kinds of experimental houses popped up—from geodesic domes and earth ship houses, to yurts and hogans. Calling themselves counterculture, these young people saw New Mexico as a place of possibility, a place to fashion new ways of living out of ancient traditions.

When La Luz's development crew first cut into the dirt on the west mesa of Albuquerque, then, they were part of this experimental and optimistic scene, building new kinds of communities in the wide-open spaces of New Mexico.

Cutting into the mesa
(Courtesy of Ray Graham)

Almost all of Albuquerque's population resided on the east side of the Rio Grande, in blocks of subdivisions imitating suburbs, in a sprawling semi-urban grid. To the west of the city, west of the Rio Grande, rose sand dunes climbing to the distinctive cones of ancient extinct volcanoes that marked the western edge of the Rio Grande Valley. The area from the river rising westward toward the volcanoes may have looked empty, but it actually bore the marks of people using this area 400-700 years ago—petroglyphs carved by native and Spanish people and the remains of many older pueblos that existed when Coronado first marched through the state in 1540. Archeologists have found not only pottery shards from these communities, but recently discovered sixteenth century metal from battles with the Spanish. While the vision for La Luz was experimental, it rested on ancient roots.

Ray Graham was one of those attracted to New Mexico. He arrived with his wife Barbara in the 1960s from Virginia, by way of a stint as Vista Volunteers in Montana. At a dinner in Corrales, Graham met Didier Raven and Steve Baer, who, like Graham, were also interested in developing innovative communities different from eastern subdivisions, something that respected the traditions and landforms of New Mexico.[2]

Raven had built a house in Corrales constructed out of terrones from Isleta

Pueblo. "It was large, rectangular, a very simple house. I and two brothers-in-law . . . laid every one of the 30,000 terrones," Raven said in an interview with the authors. Raven had earlier befriended a young architect, Antoine (Tony) Predock. Predock, he said, would come up and stare at Raven's house while it was under construction. "A little later I helped Steve Baer," a local inventor of Zomes, "build his first dome. The first whole earth catalogue with Stewart Brand was actually produced on the floor of the Baer house." Raven went to California for a year, and when he returned he met Graham at the Corrales dinner party where they started talking about doing an innovative housing project.[3]

Graham provided the capital, and Raven scouted for land to purchase and also recommended Predock as architect. "I had seen the law center at UNM (that Predock had designed). I took Ray to see it. The coffered ceilings caught our eye." Land was purchased from Joel Taylor, whose name is reflected in the large west-side development, Taylor Ranch.[4]

In the mid 1960s, Antoine Predock had returned to New Mexico after getting his architecture degree from Columbia University, touring Europe for a year on a fellowship, and working with architecture firms in New York, Massachusetts and San Francisco. He had first come to New Mexico as a nineteen-year-old student attending the University of New Mexico's engineering program. At the University of New Mexico Predock took a course in drafting taught by architect Don Schlegel, later the Dean of the fledgling School of Architecture. Schlegel became a mentor, encouraging Predock to study architecture.

In an interview with the authors, Predock told us that his father had wanted him to study engineering, but, he said, putting his hand to his heart, engineering was too hollow for his sense of aesthetics. He wanted more of a heartfelt profession.[5]

When Graham conceived his idea of a new urban project, creating sustainable housing in an environmentally friendly planned area development and commissioned Predock to design it, he was recruiting a young architect with both a regional and modern vision, one who knew about and respected the indigenous history of architecture in New Mexico. According to his biographer Christopher Curtis Mead, Predock rejects imitation but acknowledges modernists Frank Lloyd Wright and Louis Kahn as important to his vision, rooted in a uniquely American modernity. Mead states that "Predock did not mean for his work in New Mexico simply to confirm precedents from the 1950s: He figured out La Luz on his own by working through the unique conditions and circumstances of New Mexico."[6]

As Predock was envisioning a regional and contemporary style of architecture for La Luz, he was quite aware that he was not working in a blank architectural landscape. There was already a rich history and legacy of building—from the ancient native people and the Spanish past, as well as from a famous New Mexico architect, John Gaw Meem, who drew his vision from both cultures.

Raised in Brazil and educated at Virginia Military Institute, Meem came to New Mexico in 1920 to recuperate from tuberculosis. He worked the rest of his life to preserve the traditional architecture of the Southwest, a blend of native and colonial architecture called the Spanish Pueblo style. Among Meem's many distinctive buildings is the University of New Mexico library, the chapel, and other buildings on campus.[7]

The University of New Mexico library (Anne Taylor, author's collection)

John Gaw Meem's entryway
(Anne Taylor, author's collection)

Meem's architecture did not conform to the contemporary modernist idiom, but he believed that ancient shapes do evoke a modern language native to the region. Upon seeing the University of New Mexico's pueblo style campus, Frank Lloyd Wright declared to Meem that all copying was base. Yet Meem contended that regionalism—the use of old forms and traditional simplicity—was legitimate. Meem's aesthetic was grounded in history and place, a foundation that according to James Moore creates "comfort in a harsh environment...a sense of rootedness, a feeling of well being that comes from the way in which we know our place in the natural and cultural landscape."[8]

Although Predock's aesthetic is modern and quite different in design and execution from Meem's work, attention to the land, its history, materials and forms has also been important for Predock's architecture.

"We're really interlopers here," Predock has said. "The fact that I'm here involves a mediating layer. I'm this gringo from nowhere, with Chaco Canyon culture out there dating from the eleventh century, with even earlier vestiges around the West, and full-blooded descendants of those cultures around me."[9]

"When I am working on projects with my team—and it is important to underscore the collaborative component in my work—we remind ourselves that we are involved in a timeless encounter with another place, not just a little piece of land" (Collins and Robbins, 13-14).

At the time of the genesis of the La Luz project, Predock was working for architect George Wright and had designed an addition to the University of New Mexico Law School. He resigned his position to come aboard the larger La Luz project. Immediately he realized in his analysis of the La Luz site that houses needed to be placed on the higher ground to experience the beautiful view of the Sandia Mountains to the east. He felt that the houses should form an "escarpment" that echoed the elevation of the distant extinct volcanoes. In drawing his plans he was greatly influenced by Ian McHarg, a Scottish landscape architect who authored the 1969 book *Design with Nature* and believed in regional planning using natural systems. These ideas coalesced into Predock's new model for urban planning that bypassed traditional developers and offered almost a spiritual connectedness to the land.[10]

Taos pueblo
(Courtesy of James C. Wilson)

La Luz
(Courtesy of Jim See)

The adobe homes of La Luz indeed impart a strong feeling of "rootedness... of well being that comes from...the natural and cultural landscape." Residents have remarked that living with mother earth, within stacked adobe bricks, feels like being embraced and rooted. A longtime resident recalls, "My kids say that they were very influenced from living here. Both (do) work in the visual arts and still have strong connections to the people they grew up with here. The architecture is like living inside sculpture, there's something grounding and powerful about it."[11]

Rina Swentzell, a Santa Clara Pueblo architect, notes that "humans and their built expressions are but one aspect of an inter-related sacred whole. That whole includes human beings in a sense of cosmic relatedness with the local environment—with the sky, mountains, wind, animals, stone, and minerals. In this kind of thinking the human-built structure is not separate from the local place, from the land. It can only be conceived in its place or in its relationship to the local natural context. Even more, the main purpose of the built environment is to express the connectedness of humans to the larger ordering of the natural/spiritual world."[12] As we explain, the architecture of La Luz connects its residents to the surrounding land and sky and to one another in a harmonious whole that is physically and spiritually sustaining.

Transplanting trees from the bosque
(Courtesy of Ray Graham)

2

Adobe Mud and River Trees: A Part of the Environment, Not Apart from It

The town houses of La Luz cluster and step up to fit the contours of the west mesa, leaving common spaces of walkways and fountains, tennis courts, and pool. Each home has two patios and a stunning private view of the mountains to the east and southeast. Garages and roads are located at the perimeter of the clustered buildings. In 1973, Graham's Ovenwest Corporation created a covenant with the La Luz Landowners Association to preserve and maintain an additional thirty acres of mesa as permanent open space stretching from the cluster houses to the bosque.

As a result La Luz faces mostly open mesa, and that land abuts the bosque in the near distance, with the mountains visible to the east. This habitat appears to be a barren high desert, but it had been used for grazing and farming and supports a wide range of native plants, such as four wing salt bush, sand sagebrush, grama grasses, jimson weed, prickly pear cactus, apache plume, winterfat, snakeweed, globemallow, desert willow, chamisa, scorpion weed, yucca, purple aster, and more. Many of these plants had been used for food, medicine, or clothing by native people. The bosque is more verdant, with cottonwoods, one seed juniper, New Mexico olive trees, poplars and aspen, bulrushes and cattails, willow, mulberry, tamarisk and more. In fact, La Luz's construction crew dug up young saplings from land owned on the bosque to plant in the development.

As Predock began to plan and design the various cluster housing groups he executed very careful studies of the site including the movement of the sun from north to south and east to west at appropriate seasonal times during the year. He also analyzed the New Mexico wind on the west side of the Rio Grande, which can sometimes be very powerful with erratic directions and speed. "We did a site analysis [using] a box of index cards noting wind direction, sun direction, demographics. It was my urban designer's version of site analysis."[13]

Predock began his design with a clay model; "my process is total play," he told us. The homes were to be built at different angles to capture the best view of the land and the city to the east, the bosque and the sacred mountain, Sandia (which means watermelon in Spanish). The site captures the movements of the sky—moonrises over the Sandia, cloud formations, and the rosy coloring of the Sandia and Manzano mountains reflected from the sun as it sets in the West. The siting of the complex was attuned to both the land and the habitat.[14] Kenneth Frampton has called such attention to siting "critical regionalism": "an architecture whose close attention to topography, context, climate, light, and tectonic form...reintegrate[s]...nature and culture by once again grounding buildings in a coherent sense of place."[15]

Predock said, "given the acreage, I thought it wrong to scatter the houses. Instead, we should concentrate development on high ground and preserve the acreage (500 acres in the original master plan). I thought we should make a community. We called it La Luz Community. I master-planned the site for commercial development too. There were originally two loops, but we only built one loop."[16]

His intention was "to preserve the site, floodplain, the bosque, and to build an escarpment of buildings not unlike the volcanic escarpment." There was to be a community building (which became the sales office). They applied for special zoning because nothing had been done like this development before in New Mexico. The special zoning "meant that we could do mixed use, quasi-private streets that were much narrower than most, without curb and gutter requirements." Predock continued, we went "up to the higher ground to facilitate the eastward sights, to chose the topography."[17]

It was Didier Raven's job, when working with Antoine, to stand looking eastward with a story pole to figure out the views of each unit. As Didier recounted, "he'd send me up around the sand. We didn't have computers; he'd move me, and he'd draw heights and elevations as he went. He had an idea in his mind almost immediately about where the development should be, [even though] the siting didn't originally seem to make sense to others. The choice was brilliant and the first drawings were brilliant."[18]

"You join the design with the site," Predock explained. "There was no green architecture then," but "I called it ecological planning. Units followed the land, flowed with the land" and were part of it, not apart from it.[19]

Units flow with the land
(Courtesy of Robert Peters)

Pouring concrete
(Courtesy of Ray Graham)

"I used modern materials; I was not doing [traditional] vigas. We used adobe in a modern way, with knife edges. Ernie Sanchez was the adobe guy. At first we made the adobes on site, bringing in caliche to mix with sand. I wanted to use concrete lintels" to support a large, "clear span for big expansive views open to the east." There were limited openings to the west, because of the low, yet strong sun to west. "Orientation was critical, and simple materials—adobes, wood used very naturally, brick floors, all local materials."[20]

A Recipe for Adobe Brick Making[21]

    New Mexico bricks are made in wood forms and are 10" x 14" x 3.5
    Provide a mixing station with a stock tank in which to mix adobe
    Add 12 gallons of water to the stock tank (always add water first)
    Make a two-foot pile of adobe dirt
    Sift the dirt through a screen to get rid of lumps
    Have two pails (3 gallons) for water
    Use rain barrel water if possible to conserve
    Add 12 gallons of water to the stock tank
    Add 12 gallons or 7 buckets of adobe dirt (is the glue)
    Add 2 buckets of chopped up straw (Pass through a leaf mulcher)
    Add 7 buckets of concrete sand
    Need 50% concrete sand and 50% adobe mix
    Mix by hand or with hoe
    Hand squeeze mud mixture and wipe around edges to remove lumps
    Transfer mixture to wheel barrel and take to the form station
    Pour mixture into forms laid on ground
    Level the mixture with a screed tool
    Sprinkle water over adobes with watering can after mud is in forms and leveled
    Drying time will vary
    After one day remove form from blocks
    Stand blocks on end and use a small hand hoe to scrape surfaces
    When totally dry stack on pallet ready for transport and use

"Tony was obsessed with giving everybody two things, privacy with a view," Didier Raven explained. "That's still one of the most difficult things to do and one of the real hallmarks of his talent. Each house had a cardboard model, and Antoine still has some of them. You could lift off the roof, with the patios and look inside the model. The three first house types had a model. Nobody had ever done a berm in New Mexico. Antoine was intent on those big cottonwoods coming in. We got them from the river."[22]

Didier continued, New Mexico had no condominium law on the books at that time. "We had to lobby the legislature to have common walls and electricity in common walls. Along the way we were fortunate to find Gunnar Dahlquist [as general contractor]. . . . Don Shlegel had spoken highly of him. He had done a Methodist church. Antoine and he got along well together. One of the first decisions we made was to make our own adobe."[23]

Building with adobe
(Courtesy of Ray Graham)

Cluster of units, perimeter roads
(Courtesy of Ray Graham)

A plaza fountain
(Cynthia Lewiecki-Wilson, author's collection)

Architect Antoine Predock's idea of attached adobe homes as "neighborhoods" supported community and sustainability. The contiguous walls of the structures made for efficient heating and cooling. Homes were placed around and near a plaza that had a circular water fountain, brick paving, and trees and landscaping. The design of these "neighborhoods" created spaces for community, social groups, and occasional concerts and picnics.

Each home is connected to the adjacent home by a sound proof party wall made of 16" adobe (two 4 x 8 x 12" adobe bricks wide) providing a thermal mass to retain the heat of the sun as well as the cool of the mountain nights. Predock also noted that the double brick adobe walls provided good sound insulation between units. Units have brick floors (providing for heat and cooling retention), large glass windows, several skylights, and open interior spaces. Homes are situated so that all have an excellent view of open desert space and mountains to the east, but they also turn inward with walled patios that act as solar collectors and as an additional room outdoors visible from most living rooms. There are variously sized town houses, all of which are designed as cluster housing to minimize their physical presence on the site.

The construction crew was made up of eight to ten people, as well as subcontractors, supervised and coordinated by Gunnar Dahlquist, the general contractor, as well as workers from San Felipe pueblo who laid the brick floors. As with any construction, land must first be cleared and leveled where necessary. Bulldozers tried not to move too much dirt thereby preserving the land and the plants of which they were a part.

Much of the construction material and landscaping materials for La Luz came from the local land and river bed. While the land was being readied for laying foundations and walls, the adobe bricks were initially made on site. Later, as more and more adobes were needed, production was moved offsite, across Coors Road, for faster and greater production.

The walls of La Luz homes are constructed of these hand-made 4 x 8 x 12 adobe bricks, stacked without grout between each unit, two wide for outside walls, then finished with chicken wire, a scratch coat, and lastly a color coat of stucco. The fireplace bricks, floor bricks and those in outdoor patios came from Kinney, a local brick maker. The fireplace exteriors were wrapped in adobe—some of these walls are curved—and then stuccoed. Canaletas (roof drains) and bond beams were made of concrete poured on site.[25]

A patio
(Courtesy of Robert Peters)

Constructing with adobe
(Courtesy of Ray Graham)

The designing and construction began at Arco site 1-9. The construction then progressed to Berm 1-20, then Link 1-24, Pool 1-19, then Tumbleweed 1-10 and Tennis Court 1-35. After all the building and before Graham moved to start building another project (La Luz del Sol across the street on the west side of Coors Road), architect Hildreth Barker of Barker Bol, Architects, was hired to design four more units (10, 11, 13, 14 on Arco), built by John I. Chavez.[26]

la Luz site map (Courtesy of La Luz Landowners Association)

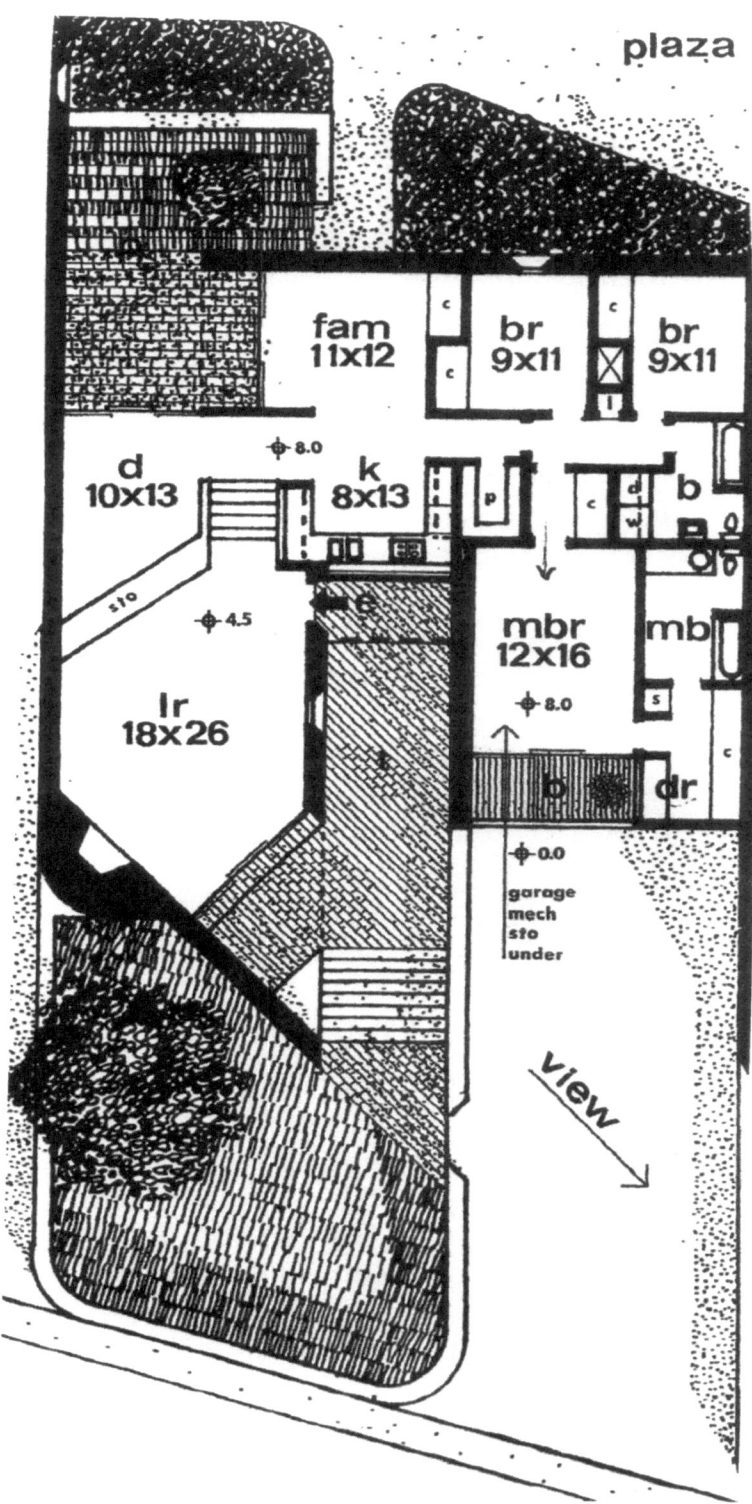

Architect's drawing (Courtesy of Antoine Predock)

There were ten different floor plans, some two bedroom, three-bedroom or four-bedroom homes with private patios. The following are a sample of six designs.

Floor plan #1: Approximately 1900 square feet, 3-4 bedrooms, and 2 baths

These houses are raised so they can look past the houses on the opposite side of the street to the bosque, the Sandia Mountains, and downtown. As one walks through the gate of one model there is a lower patio with brick steps leading to an upper patio and the front door. One enters into a large sunken living room, which leads up more steps to a dining and family room and a kitchen to the side of the steps. A side hall leads to the master bedroom with attached bathroom, and a balcony over the driveway. There are two (or three) other bedrooms and a bath. Some similar homes have a basement with a spiral stair leading to the below level area.

Floor plan #2: Approximately 2,000 square feet, 4 bedrooms, and 2 baths

In this model the entry is through a courtyard leading to a foyer, leading on the left to the living room with a dining-L and kitchen with an outdoor patio off the dining room that beckons one to a large south yard and covered patio. There is a hallway of a few steps leading to the master bedroom and a bath. From the steps one can look at a high rectilinear window in the living room and see the sky and the Sandia Mountains. There are three other bedrooms and two baths.

Floor plan #3: Approximately 1826 square feet, 3 bedrooms, 2 baths

This model has a small entryway through a courtyard. Other units have their entry from the street through a larger courtyard. Some have attached garages. Others have detached garages. The entry leads to a kitchen with a counter that separates the living and dining areas, and a view through the dining room and sunken living room to an east facing walled patio and the Sandia Mountains. A side hallway leads to a master bedroom and bath on the east view, and two bedrooms and bath on west end of the hallway.

Interior view
(Courtesy of Jim See)

Floor plan #4: Approximately 1900 square feet, 3 bedrooms, and 2 baths

The courtyard entry is designed so that it looks into and through the kitchen windows on the east side, straight to open mesa, the bosque and the Sandia Mountains. Anyone approaching through the courtyard thus shares the views, as if looking at a painting. The entry leads to into a side entry foyer, and then into a large dining room. Steps lead from the dining-entry into a sunken living room and an outdoor walled patio on the east facing the mountains, which acts as another room in the summer. To the west of the entry are a study, now a music room, and an office (former bedroom). A hall beyond the dining room leads to a master bedroom, dressing room and bath. In the original plan there are two additional bedrooms down the hallway, but because of modification to create a music room and office, there is presently one additional bedroom and a bathroom.

Floor plan #5: Approximately 1550 square feet, 2 bedrooms, 2 bath

Entering from the street, an attached garage forms a gated courtyard leading to the front door. To one side of the entry is a kitchen and to the other is a hallway with bathroom, and stairs leading up to the second level. In front of the kitchen on the first raised level is a dining room looking eastward to the mountains. To the side of the dining room are steps leading down to a rectangular living room, which opens onto a patio on the east side. Upstairs are a master bedroom with fireplace, and a guest bedroom with a hall bathroom.

Floor plan #6: Approximately 2300 square feet, 3-4 bedrooms, and 2 or 2 1/2 half baths.

Entering from the street, an attached garage forms a gated courtyard leading to the front door. An entry hallway stretches from the front door through the lower floor of the house and extends down three steps to a doublewide living room with fireplace, flanked by large windows on each side. One or two sliding doors lead to an east-facing patio. Off the entrance hallway on one side is a kitchen with a pass through counter to a dining room east of it, overlooking the living room. To the other side of the hallway is a den with adjoining half bath. In some units this space is for washer/dryer and hot water heater instead of a downstairs bathroom. Across from the dining room, stairs ascend to the second floor and an upper hallway. To the east, the master bedroom with fireplace extends the width of the house with views of the mountains and downtown Albuquerque. A master bathroom is off this bedroom. Two additional bedrooms and a second hallway bathroom complete the upper level.

A unit on Berm Street (Courtesy of James C. Wilson)

Plaza fountain looking toward mountains
(Courtesy of James C. Wilson)

Over the years, La Luz has won numerous accolades from architectural organizations and national and international magazines. Almost immediately it garnered attention for its "creative conservation, architecture and planning." It was, according to *Architectural Forum*, "a show of courage."[27]

Reflecting back over fifty years on his first architectural commission, Antoine Predock told us, "my passion for La Luz revives when I walk its different wandering paths. It is not a grid. There's a standoff with mesa, where it meets the cultivated areas. It has this aura of New Mexico but in a different way...[there are] observatory channels and views. It is not directly influenced by Chaco but [evokes] the aura and power of place, the cultural layers strata, for example, Pueblo Bonito's highly organized markers, such as the straight wall."[28] "There is a deep spiritual sense in New Mexico of the land and what the early cultures left us."[29]

Pueblo Bonito, Chaco Canyon
(Courtesy of James C. Wilson)

Memorial Day picnic 2018
(Courtesy of James C. Wilson)

3

Community: An Experiment in Democracy

A visitor to La Luz in May might hear the soothing plash of the fountains, smell fresh cut grass or the aroma of baking cookies drifting through an open window. Our visitor might hear young children attempting to play bocce ball next to the pool, or pickle ball with paddles on the tennis courts, or maybe the children are riding bikes on the safe and semi-private roads. Mom and Dad might be following to demonstrate how to throw the bocce ball underhand and how to score a game, or perhaps to supervise them in the swimming pool, or maybe to play their own game of tennis nearby.

Our visitor will no doubt hear a summer medley: children at the pool screaming with delight, the "wak, wak" of a tennis ball, the calls of residents as they gather for the annual Memorial Day picnic near the tree in the green grassy meadow below the pool. On this day there may be a book and jewelry exchange on tables under one of the shady trees. Some older residents might be seen playing a game of croquet. Add to this scene the smell of grilling hot dogs and hamburgers; the sight of tables overflowing with side dishes and sweets, fruit and wine; and you have a picture of the holiday life, an important part of the community spirit at La Luz. Such community festivities take place in May, July, September and December every year with traditions that give La Luz the feel of a small-town community.

Memorial Day picnic
(Courtesy of Helen Reilly)

Fourth of July picnic
(Courtesy of Jennifer Fenstermacher)

While La Luz exists inside the bounds of a large city within a five-mile radius of the city center, its architecture and siting contribute to this feeling of a small town. Homes are grouped together, with common "party" walls and shared public walkways and plazas. They are surrounded by a larger preserved open space, with bosque, river and mountains in view. Every home has a unique view. The special quality of La Luz is rooted in its design and in the relationships between land and people, sky and mountains, history and the community that its architecture fosters.

Community was foremost on the minds of Ray Graham, Antoine Predock, and Didier Raven, who were in their late twenties and early thirties when they embarked on their project.

"I had read material from the Urban Institute in Washington, about housing in Nordic countries, Finland," Raven explained. "We wanted not to have a high income development. We were in our late twenties, Antoine just thirty. Our whole idea was to make it a moderate income [development] with a lot of children, small streets, and no cars so the children could go to the pool, walk around without danger. In the first thirteen houses that were built, there were five children, including my two."[30]

Antoine Predock's parents also lived in their own unit.[31] This feature, several generations living side-by-side, is another element that gives La Luz its small-town feel. A current resident told of her parents retiring to La Luz after living abroad. She initially moved into La Luz to be near them as they aged. Later, her daughter and grandchildren lived here.[32] Other residents have replicated this pattern, with elderly parents and their adult children and grandchildren living in different units of La Luz. Some children who grew up in La Luz have returned to live here as adults. Another resident fondly recalls playing at La Luz as a child when he visited his mother who was then a resident. Many years later as an adult he moved here, and his son was born at La Luz, in perhaps its only home birth.[33] One family moved in when their unit was first built; they later moved away to another part of town, and later still returned to live at La Luz. Their daughter spent her teenage years in La Luz, and she later returned to live here with her partner, who together raised their daughter. This resident said she appreciated the ease and safety of living at La Luz with a child, who could walk across the traffic-free plazas to see her grandparents. Like several other La Luz children, she attends the Bosque School, which borders La Luz on the east side.[34]

Space for a school was in Ray Graham's original plan for La Luz, though one was only built several decades later. The Bosque School bought its present site and campus of twenty-three acres adjacent to the bosque at Coors and Montaño in 1999. A college preparatory private school founded in 1994 for students in grades 6-12, the Bosque School offers a special emphasis on Environmental Studies. An early program was the Bosque Ecosystem Monitoring program (BEMP), a joint effort of the University of New Mexico Biology department, the Bosque School, and thirty-two research sites along the Rio Grande River, extending from the Ohkay Owingeh Pueblo north of Santa Fe to the Mesilla Valley State Park near Las Cruces. Students and volunteers conduct research along the Middle Rio Grande and its associated forest, collecting data related to the overall condition of the forest ecosystem and the Rio Grande and reporting their findings to the University of New Mexico. This program is now a part of The Albert J. and Mary Jane Black Institute for Environmental Studies, a center for bosque riparian, and watershed research.[35]

Budagher Hall, Bosque School 2018
(Courtesy of Elliot Madriss)

Quad from Budagher Hall
(Courtesy of Elliot Madriss)

When the Bosque School was raising money to purchase additional access land (Tract 4) several residents of La Luz dug deep into their pockets. Their donations totaling more than $400,000 helped to preserve open space and the sweeping views.[36] Today there is a permanent marker along Learning Road thanking these generous and far-sighted La Luz donors. The Bosque School also owns and maintains a residential unit in La Luz. When a La Luz resident and architect was hired by the city's public arts program to study and create placards of the animals, birds, trees, and river for the Montaño bridge, he enlisted the help of Bosque students and a teacher, who collected data for the project. For all these reasons, La Luz and the Bosque School share a special relationship.

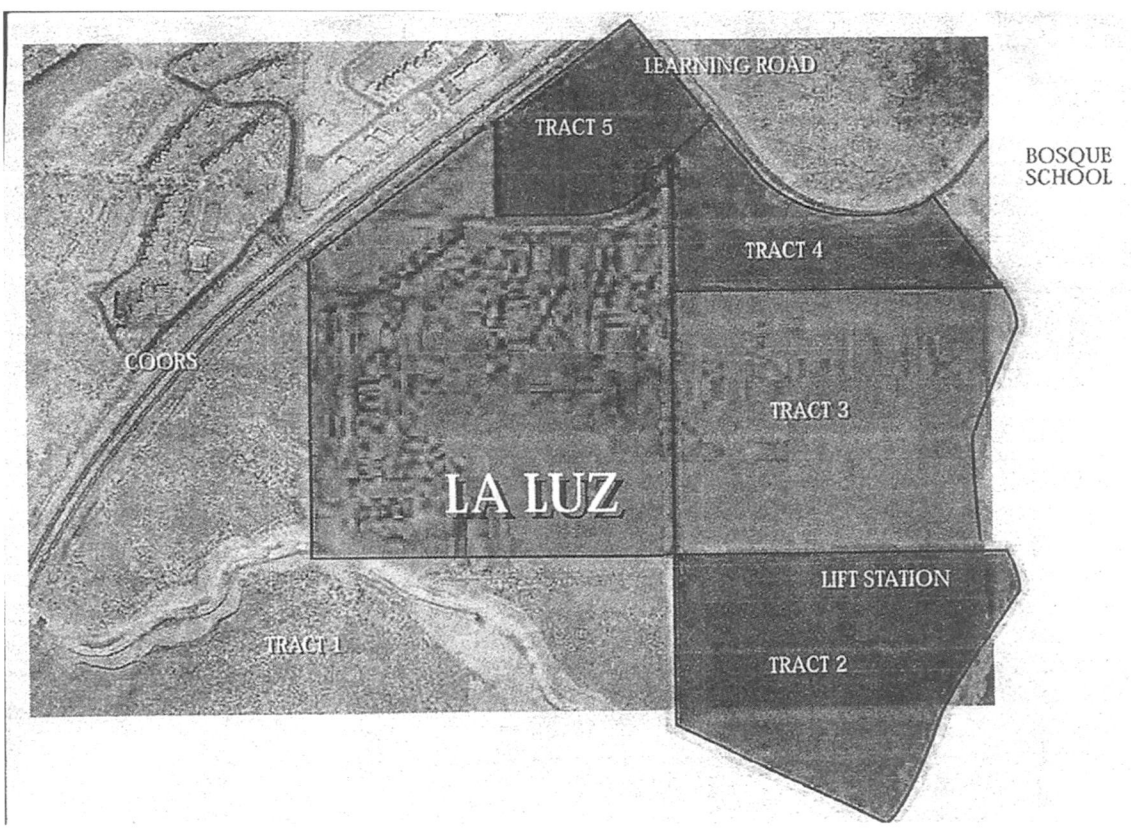

Tract map
(Courtesy of La Luz Landowners Association)

In 1969, Ray Graham's Ovenwest Corporation, established the La Luz Landowners Association to "take title to common areas and facilities" and "to administer" them for the good of the community. Ray A. Graham, III, Didier Raven, and Graham Browne comprised the first Board of Directors.[37] This board established community-wide regulations, such as controls on signage, traffic, and maintenance. In 1973, Graham conveyed a deed for Tract 1 to the La Luz Landowners Association and its Board of Directors, now composed of elected residents who took over governance of the community. In 1998 new bylaws were enacted, formalizing voting rights (every unit has one vote), assessments, common land use, special charges, reserves, architectural standards, maintenance, and other responsibilities and rights.

Over the years, the Landowners Association has become central to the democratic governance of La Luz and acts as a social catalyst, giving residents the opportunity to express their interests and values within a compatible group. Participating in self-governance contributes to a lively social environment and gives residents a strong sense of identification and self-determination as a community.

The model of La Luz's self-governance is based on representative democracy, and is a microcosm of city, state, and federal governments. There is an annual meeting each year. Before the meeting, residents vote by paper ballot on members to serve on the Board of Directors. Those elected then choose a President, Vice President, Treasurer, and Secretary. At the annual meeting or thereafter, residents can voluntarily sign up to serve on Committees. The president appoints Board members to serve as liaisons to each committee, and committee chairs often attend board meetings. Some of these committees were established early, in the 1970s, and some have evolved over time as needs and interests became apparent. The committees include Architecture, Common Grounds, External Affairs, Finance, Maintenance, Nominating, Publicity, and Social (formerly called Recreation).

Architecture

The primary responsibility of the Architecture Committee is to advise and to assist the board of the La Luz Landowners Association in preserving the architectural integrity of La Luz as originally designed. It has established architectural Standards and Procedures that must be followed. The committee informs landowners of these standards, which help maintain the exterior appearances and value of the property, and its integrity as a historic and cultural property.

External Affairs

The External Affairs Committee works with the surrounding neighborhoods, the West Side Coalition, and the Albuquerque coalition of neighborhoods to monitor surrounding development, preserve the integrity of views, and lobby for sensible land use. La Luz was one of the founding organizations of the West Side Coalition, and it has been very active in arguing for good development and preventing big box construction in the immediate area. The chair of External Affairs is notified of meetings about building and zoning changes in the surrounding areas and attends many city and area meetings, and then reports back to the La Luz constituency.

Finance Committee

The Finance Committee oversees budget planning and oversight.

Landscape Committee

The Landscape Committee was originally part of the Architecture Committee, but became a separate committee in the 1990s. It works to preserve the architectural integrity and design of La Luz, but also to allow room for individual residents' preferences. The committee monitors landscape concerns, such as trees blocking views, plant health, and community gardens. According to a long time member, as early as the 1990s the committee became concerned about water scarcity and started to educate the community about the need to conserve water.[38] This committee has come into greater prominence in recent years as La Luz changes its landscape to one more compatible with water scarcity. As a result, the Common Grounds Ad Hoc Committee was formed to implement a new landscape, hiring landscape architect Richard Borkovetz to create a plan that features native plants and grasses. Over time, the original bluegrass is being converted to less water dependent plantings.

Maintenance Committee

The primary responsibility of the Maintenance Committee is to advise and to assist the Board of Directors of the La Luz Landowners Association and to provide direction to the Manager in preserving and maintaining the physical environment of the common areas and facilities of La Luz. This committee provides prioritized plans for safety and security, repairing damage to the property, and performing periodic inspections of the infrastructure. It works with the Landscape Committee to provide plans for planting flowers, shrubs and trees in approved areas, and writing and reviewing a yearly maintenance plan and budget for submission to the Board. This committee has assisted in the writing of the *Handbook for La Luz Landowners*.

Nominating Committee

The nominating committee meets at least once a year to form a ballot for the selection of the next year's Board of Directors. After asking the community for volunteers who wish to serve on the Board, and soliciting names from residents, the Nominating Committee determines a slate and mails ballots to each La Luz unit landowner with the candidates' names and biographical data.

Publicity Committee

The publicity committee is responsible for publishing the community newsletter, *The La Luz Voice*, four times a year. The newsletter features profiles of new residents, news and pictures of recent community events, and occasionally reprints some historical articles about La Luz.

Recreation (Social Committee)

The primary responsibility of the Recreation Committee is to plan a recreational program geared to the interests and needs of the residents of La Luz. Consideration is given to social, cultural and educational programs and for ways to determine the kinds of programs that interest residents. This has resulted in potluck picnics in the meadow for Memorial Day, Fourth of July, Labor Day, book and jewelry exchanges, musical concerts, and occasional movies and lectures.

Over the years this self-governance structure has remained fairly stable, with some variations. In 1974 a permanent manager position was created. The issues the community debated in the early to mid seventies were very similar to today's issues. One innovative idea of 1976 that has relevance today was the block meeting. Each neighborhood would meet four times a year, with a board member attending, to discuss concerns, promoting governance from the bottom up.[39] Issues facing community debate are recorded in minutes from the meetings of the seventies, eighties, and beyond. These issues include roads, office and storage space, landscaping, security, the swimming pool, trees and view preservation, water conservation, maintenance of sidewalks and plazas, traffic and parking, and of course the budget and annual dues. Challenges in those earlier years seem to be the same as those the community faces today.

Some Early La Luz Residents

One of the early Presidents of the Board was a former architect for Skidmore, Owings and Merrill living in Chicago. He often visited a friend and architect in Taos in the late 1960s and early 1970s and visited La Luz a few times while it was under construction. He fell in love with the light and shadows of La Luz, the effect of adobe—its solidity and mass. He was interested in the fact that La Luz made use of traditional materials, with the addition of modern concrete. He also was intrigued that La Luz was designed as a system—the berms, the pedestrian walkways, the cluster housing, variations in floor plans, recessed doors, the height of patio walls preserving privacy, and generous open space per home. In 1974 he purchased a home at La Luz after completing his own site analysis. He photographed all the views from each house, laid them out on a table, decided which had the view he wanted, and then purchased that home with the best view.[40]

Like other residents, this architect moved to La Luz from out of state, attracted to the pristine beauty of New Mexico's landscape and architecture. In recent years, homeowners have come from Washington State, California, Texas, Ohio, Washington, DC, Denver, Massachusetts, and Australia by way of New York City.

This early resident has served on committees during all the years he has lived at La Luz, three terms on the Board and on the Architecture committee, and chairing each. In the early years, he said there were fewer residents, and people had to learn how to run a landowners association and live communally. Reflecting back on his forty-five years living at La Luz, he believes the vitality of the community has been preserved through remodeling and upgrading of individual homes and older systems of the complex.[41]

A President of the Board in the late 1990s recalled one of the issues the community faced in that decade "was the easement we granted to the water authority for the arroyo to the south. We owned part of it, and the water authority wanted the easement to maintain it." This resident has been active in governance over the years, serving on the Finance committee as well. He noted "a movement toward professionalism—the board is more interested in the details of running La Luz," he said. "One of the nice things about the committee structure is that people gravitate to those areas in which they are interested. The committee structure is one of our strengths here. Some HOAs turn over entire management to a management company, and then people lose the feeling that they have a voice. The governance becomes less a democracy in action."[42]

Another past Board President of the early 2000s recounted the community's efforts to fight big box developers from building in adjacent tracts. She noted that she "did a couple of tours on the Board" and mentioned that an important accomplishment "was to establish a retirement fund for the manager that hadn't existed before."[43]

Another Board President affirmed that working with other neighborhood associations to fight big box developers was a high mark of La Luz community efforts when she was President. She also became involved with the Community Associations Institute and worked with others on the reserve fund.[44]

A resident who chaired the Board twice and served on the External Affairs committee described the construction of the berm along Coors Road to protect La Luz and especially residents of Link Street from traffic noise. He recalled that when Coors Road was to be widened to six lanes, residents went to the city to talk about the project. "We said, build us a berm.... The city showed up one day and talked, and couple days later brought out earthmovers and in a couple days had the berm built. It was vast and barren and looked terrible. For the first years, only tumbleweeds grew. We kept pulling out tumbleweeds, but [we eventually learned that] the tumbleweeds play a role in other plants growing. Then we put in irrigation along Link and planted native grasses and plants. After the traffic on Coors Road increased, the efficacy of the berm was shown with reduced sound readings of 5-10 decibels."[45]

In addition to the overall governance structure, with its Board and standing committees, residents have long formed into small social groups of people with shared interests. There are several active book clubs; some people even belong to more than one book group. Members generally meet once a month in someone's home, and a volunteer leads discussion on the chosen book. As befitting of dedicated readers, there are many books to share and a plan to convert old mailboxes to little free libraries. Many residents are also published authors—from retired university professors and other professionals who wrote in their careers, to travel writers, essayists, novelists, and poets.

Many residents are avid bridge players, with four to six people meeting monthly in someone's home. Games are generally informal, with players rotating in and out of a seat at the table, depending on the number of attendees. Like the reading groups, some bridge players belong to more than one group and meet several times a month.

Other traditions that have been popular in times past were tennis tournaments on La Luz's courts, bocce ball on the recently built court behind the pool, Christmas caroling, and the setting out of luminarias during the Christmas holidays. Music also plays an important role in the community. Small groups of residents meet to make music together, and every year the community hosts one or two classical, jazz or semi-rock concerts on its plazas.

Jazz concert on a plaza
(Courtesy of Cynthia Lewiecki-Wilson, author's collection)

In addition to parties, potlucks and shared community holiday celebrations, many residents have long enjoyed bird watching. One resident who has lived at La Luz for forty some years has compiled dozens of notebooks cataloging the birds she has seen on and around the open space of La Luz. She and a more recent resident shared a list of some of their bird sightings: egrets, western king birds, swallows, black chinned hummingbirds, broadtails, several kinds of woodpeckers and warblers, pheasants, white breasted nuthatches, grackles, mountain chickadees, black headed grosbeaks, great horned owls, killdeer, Canada geese, dark eyed juncos, flickers, several kinds of black birds and hawks, northern harriers, thrashers, turkey vultures, meadowlarks, kestrels, quails, roadrunners, phoebes, bushtits, brown towhees, loggerhead shrikes, northern shovelers, ruby-crowned kinglets, and more.[46] One of the most exciting times at La Luz occurs in mid November when the sandhill cranes, honking high in the sky, migrate to their Bosque Del Apache bird refuge near Socorro, or leave in the spring to return north.

Double rainbow over La Luz
(Courtesy of James C. Wilson)

The open spaces, walking trails, nearby bosque, and city bike trail across the Rio Grande Bridge on Montaño Road and along the river provide opportunities not only for birders, but also for daily walkers and joggers and bike enthusiasts. Given the natural habitat and spectacular views of land, mountains, and sky around La Luz, it is no surprise that many residents are photographers. Most of the photographs in this book come from current La Luz residents.

Storm on La Luz skyline
(Courtesy of Helen Reilly)

Clouds and balloon over La Luz (Courtesy of Helen Reilly)

A summer day (Anne Taylor, author's collection)

Whether sky and balloon watching, swimming, playing tennis, or walking, residents of La Luz enjoy a serene and healthy environment. Young children, middle-aged parents, new retirees, and many residents in their eighties and nineties mingle side-by-side, and that might be a secret to the local longevity and aging in place.

One social group that has formed in the last few years is called "La Luz at Home." Originally it was a more formal group that contracted with a geriatric management company to negotiate medical needs and coordinate volunteer care helpers. "La Luz at Home" has evolved into a more informal group that supports volunteer activities that help residents live healthy and productive lives. This group provides a variety of health, wellness, educational, and recreational activities and mutual support for each other to help maintain, health, vitality and independence. The group organizes trips to art galleries, movie outings, and summer water aerobics at the La Luz pool, along with assistance in shopping, rides to appointments and attention to other necessities of life that can make living at La Luz a lifetime experience. Various models for aging in place have emerged at La Luz.

The Single Person Model, with or without a Pet

Many people chose to live alone as members of this vital community. Some have companion dogs or cats, turtles, chickens, or other animal friends; some have no pets. Many are busy writing poetry or books, playing music, traveling, playing golf or tennis, walking, swimming, gardening, or bicycling. Some volunteer. Because of its aesthetics, La Luz fosters wellness—physically, mentally and spiritually. Like the life of the trees in the bosque, we benefit from a healthy communal support system that is real democracy in action.

The Family Model

Many residents of La Luz are fortunate to have families with children here, and those without children enjoy the sound of children's play and feel the joy they bring. Some of the children will water plants for older residents, or walk their dogs and feed their cats while they are away. The mix of youth and older folks makes La Luz a real community.

The Sibling Model

After one resident became a widow, her sister visited frequently. Sharing time together and renewing their common interests became such a pleasant experience that they decided to live together as they had as children, even though both are now grandmothers. One is a fiddler, the other a guitar player. Both have a penchant for learning new music and modes of music and practicing. They have attended master workshops at Ghost Ranch and now play blue grass and sing. It is wonderful to see two people grow with new accomplishments. In the process they may have invented a new form of a two-person family with a rich and full and vibrant life.

Commuting Residents

Some live here with a family or as a couple, but may have another home in another geographic area. One couple spends half the year here, and half in New York. Others make their La Luz home a vacation spot, living elsewhere most of the year. The La Luz community welcomes and nourishes all these various living arrangements.

The following profiles of some of La Luz's long time residents demonstrate how those in their eighties and beyond flourish in this community.

Resident with her dog
(Courtesy of Jonathan Abdalla)

In her late eighties, this long time resident of La Luz is often seen walking her dog on our beautiful campus, which, she says, is the way she gets her exercise. Her dog came from a program in which women prisoners train dogs. This resident remembers in earlier days hiking the Oxbow (of the Rio Grande) with her children and grandchildren. They would see beavers, coyotes and rattlesnakes. One time she found the skin of a rattler with the rattles still on it. She thought it was beautiful and so brought it home, rolled it up, and sent it to her ten-year-old grandson in New York. Her now deceased husband, she noted, was President of the La Luz board at the time when money was raised for the Bosque School land purchase.[47]

Also in her late eighties, another resident has lived at La Luz for forty-five years. Now retired, she was one of Albuquerque's top public school principals for elementary and middle schools. She grew up on a cattle ranch in Texas, where her parents homesteaded two sections of land near the Brazos River. She still visits the family land and until recently still rode horseback. After her first husband died, she remarried. They were looking for a contemporary home and discovered La Luz while it was under construction. Fascinated, she drove on the dirt roads leading to the construction site, and spent many hours watching La Luz being built, so much so, that the workmen, whom she befriended, called her "the supervisor." "Living at La Luz has been a wonderful journey," she said. She has kept files of community governance, helpful to this book. It was her suggestion that we resurrect the block meetings to encourage grass roots communication.[48]

Picuris Pueblo Dancer (Courtesy of James C. Wilson)

Picuris Pueblo blessing
(Courtesy of Carol Bennett)

Community blessing dance, Memorial Day
(Courtesy of James C. Wilson)

On its twenty-fifth anniversary and again on its fiftieth anniversary, the community invited Native American Indian dancers to bless La Luz. The community gave thanks for the shared space, serene architecture, and the surrounding land and sky—all of which contribute to the vitality and health of La Luz residents. In turn, the community is looking ahead to sustain La Luz for another fifty years, a project we turn to in the last chapter.

Traditional bluegrass landscape
(Courtesy of Jim See)

New landscape of native grasses and plants
(Courtesy of James C. Wilson)

4

Sustaining the Light on the Mesa

In the study of human factors, researchers have found that the physical environment in which we live and work deeply affects our health and behavior. As Albuquerque commerce and development closes in on La Luz, our place in the sun becomes more precious every day. Because a rich treasure of a distant river, the still wild and verdant bosque, and the open space that preserves our view to the east surrounds us, we have a mission to insure the preservation and conservation of our site and its surroundings. What does the future hold and what can we do or envision as residents about our mission to stabilize, enrich, and sustain our environment?

How to sustain La Luz for the next fifty years is the first challenge. La Luz has thus far maintained a simplicity that was not perhaps anticipated. Graham's initial plan was much more complex, with two loops of housing, stores, and homes above businesses. But the community that resulted was smaller and simpler, with fewer than 100 clustered units, shared plazas, and small grassy areas with trees in between. The open tracts abutting the perimeter were not commercially developed until the middle of the first decade of the 2000s. Over the course of the half-century of its existence, the La Luz community has grown to treasure this open and austere setting. The community has worked hard to fight the encroachment of big developers and to preserve the precious views. Sustaining La Luz for the next fifty years means preserving its simplicity, while facing the challenges of ever more expensive and rationed resources, such as water, electricity, and vegetation, and continuing efforts to influence surrounding development so that it is compatible with the evolution of La Luz as a pristine jewel on the mesa.

The protection of the ecosystem in an arid climate amid global warming has been foremost on the community's agenda. The original lush gardens and grassy spaces between units and plazas clearly could not be sustained in the changing desert environment. In 2014, a new landscape plan that conserves water by using native plants and grasses was commissioned. Landscape Architect Richard Borkovetz created a plan that was accepted by the Board of Directors. Accordingly, traditional lawns and plantings are being gradually replaced with more heat tolerant and arid loving native plants. Their growth takes time and does change the look of La Luz (see photos above). Once lush green spaces now appear sparser, with the browns and greys of gamagrass and desert plants.

Landscape Master Plan
(Courtesy of La Luz Landowners Association)

The translation from graphic images and ideas of a landscape architect, as in the Landscape Master Plan, to the reality on the ground has been controversial. Starting new xeriscaping takes time; plants develop more slowly than water intensive turf grass. But a new aesthetic can emerge, one that connects the plaza spaces of La Luz to the mesa beyond.

As befits a robust democracy, in addition to the ongoing debates about the landscape conversions, there is an even more heated debate about the use of herbicides. The Board of Directors has voted to continue the landscape conversion and minimize as much as possible the use of chemicals in landscape upkeep.

La Luz has long had small community gardens: a vegetable garden (previously, that spot featured a flower garden for cutting), an herb garden for common use, a memorial rose garden, and a fruit orchard. Community members tend these gardens and use natural compost and ban pesticides. Anyone who lives in La Luz may join the vegetable brigade, but the residents who plant and tend the garden have the right to its produce—typically, tomatoes, green beans, squash, beets, chard, arugula, strawberries, eggplant, cucumbers and peppers. A resident architect designed the fence around the garden to keep out local critters such as rabbits. The herb garden features rosemary, mint, oregano, fennel, sage, thyme, lavender, chives, basil, and winter savory, as well as raspberries, with some flowers mixed in. All residents are welcome to pick the herbs, as well as fruit like apricots and plums from the orchard. The memorial rose garden established in the early 1990s was planted in memory of La Luz residents who died. As one resident said, "we did not bury anything or consecrate or dedicate the site" with a plaque, but it nevertheless serves as a "contemplative spot" to remember loved ones.[49]

La Luz Vegetable Garden (Courtesy of James C. Wilson)

Having a voice in shaping the growth around La Luz remains a central concern. Starting in the early 2000s, External Affairs committee members, in conjunction with other Westside Albuquerque groups, fought to keep big box developers from building on Tract 6 (approximately 70 acres) running from Learning Road (now the Bosque School Road) to Montaño Road on the north. In 2017 a shopping center of small stores was built on part of that land.

La Luz as a Learning Environment

In the midst of such development, La Luz's protected mesa on the east seems ever more precious. Bird watchers and other residents keep an eye out for animals that share La Luz's habitat. Recent sightings include bobcats, squirrels, porcupines, coyotes, jackrabbits and cottontail rabbits, muskrats and beavers, gopher snakes and rattlesnakes, chipmunks, voles and moles, lizards, frogs, bats, raccoons, weasels, and butterflies. As land surrounding La Luz becomes more developed, La Luz's thirty acres of protected open space is a precious habitat for birds and animals and humans learning how to live with and preserve the environment.

Cows once grazed on this space. At that time this land, now designated city Open Space, was zoned for agricultural use. While the community treasures and wishes to protect the open space once filled with prolific grass, the now unused land has become fallow and filled mostly with tumbleweeds and woody bushes, crisscrossed by a few sand trails. The feet of animals no longer crunch the barren soil of the open space. Land needs to be turned over. Judicious rotation of small animals, like goats, might help improve the soil.

Coyote beside wall of La Luz
(Courtesy of Carol Bennett)

The idea of returning some of the open land to grazing and the ideas that follow are not the products of the community's long range planning, but ones the authors have thought about. We include them here to promote discussion about the future.

One vision for the future could be to develop a larger garden area or even a greenhouse on some of this space, a nursery to start seedlings for La Luz's landscape renovation. An urban farm might even be a possibility in the future where we could have our shared "farm to table" fare.

One landscape architect visitor to La Luz estimated that we presently lose over 900,000 gallons of water off the flat roofs or our houses every year.[50] Water on our present site might be collected through cisterns dug into the ground around the sloping community, and rain barrels distributed to residents could catch water flowing from roof edge caneletas.

Water from a caneleta
(Courtesy of James C. Wilson)

The bosque adjacent to La Luz's open space is a high mesa riparian ecosystem, an important habitat for an array of species, such as beaver, muskrat, turtles, opossum, frogs, reptiles, rabbits, raccoons, fish, and birds.

Bosque School students studying the riparian ecosystem
(Courtesy of Elliot Madriss 2018)

BEMP seventh grade science
(Courtesy of Elliot Madriss)

According to the United States Department of Agriculture Natural Resources Conservation Service, "in the western United States, riparian areas comprise less than one percent of the land area, but they are among the most productive and valuable natural resources."[51] This ecosystem is a close and valued neighbor of La Luz, a place that residents visit, walk and bike through every day. To preserve this resource for the next fifty years, La Luz could partner with the Bosque School, city, and county to preserve his watershed ecosystem.

A child climbing zome
(Courtesy of Anne Fitzpatrick)

A child hanging from zome
(Cynthia Lewiecki-Wilson, author's collection)

Zonohedron close up
(Cynthia Lewiecki-Wilson,
author's collection)

If you visit La Luz on a summer day, you might see children climbing or hanging upside down from a futuristic structure near the pool and tennis court. This unique playground structure is made of ordinary steel connected in patterns called Zonohedra, which are non uniform rhombiscosidodecahedra.[52] A resident who is an engineer explains, the zome is "a substantial three dimensional structure woven from just two simple two-dimensional parts (and lots of them). [If you] lie on your back in the grass below the climber with your head in the center of the dome, looking upward you will see a uniform crystal-like pattern that is the Zome."

This zome climber, like a few others in New Mexico playgrounds, was designed and built by Steve Baer. Baer is a mathematician and inventor, not only of climbing structures and other complex geometrical houses made from old car metal, but also of toys, solar devices and solar driven machines, the beadwall greenhouse, and other kinds of technologies for people who wish to go off grid.

Baer has lived in Corrales, New Mexico, for over fifty years, and was friends with Didier Raven at the time when La Luz was just an idea. He was part of that wave of innovation in counterculture living that swept through the country in the late 1960s. Baer and his wife Holly bought Raven's terrones house (mentioned in Chapter 2). "It was beautiful," Baer told us. "It exhibited a fanaticism for something that works." Baer himself has been dedicated to making things that work, that use nature sensibly and sustainably. Baer was not involved in the building of La Luz, other than to once or twice repair a broken machine, as he recalled. But his life-long philosophy of making self-sustaining objects resonates deeply with the La Luz project.[53]

Several La Luz home owners have installed individual solar collector systems on the their homes. But a greater commitment to sustainability, in the spirit of Baer's Zomeworks, would be to develop a solar micro grid for the community. With ample gently sloping land facing east, south and west, there is plenty of sunshine to tap.

The sustainability of the ecosystem was initially part of the vision for La Luz. As architecture historians noted, "The aspiration toward the making of a timeless architecture should also contain aspiration toward social responsibility. La Luz Community...demonstrates a reaffirmation of the large-scale use of adobe brick in the context of a project whose five-hundred acre site plan was based on sustainability and protection of the ecosystem."[54]

We have collected some other ideas for sustaining La Luz in the future and enhancing this community:

We could be a living extension of the nearby Rio Grande Nature Center, accessible to La Luz residents by crossing the Montaño Bridge and heading south on the nature/bike trail. We could collect data about natural habitats of the local fauna, as they seem to move closer and closer to La Luz because of decreasing available land. We could join some of their environmental education classes on the natural riverside environment that deeply affects us and learn more about how we can help their research.

We could collect seeds and contribute to a national or international seed bank.

We could continue to develop plantings to keep a healthy "green index." A healthy landscape affects the well being of our people, can sharpen focus, reduce stress and contribute to a variety of health benefits and longevity for our residents. The La Luz campus is a healthy one and provides many opportunities for safe dog walking, bicycling, tennis, swimming, gardening, quiet yoga or meditation. We must keep and enhance these benefits in the future.

We could also contribute to fire prevention in a cohesive effort with the Bosque School and the city by helping to clean up the bosque and to keep our houses clear of brush and overgrowth of desert plantings.

Some of our concrete surfaces could be replaced by permeable materials thus saving ground water and providing a series of conservation pathways more beautiful and sustainable than cracks in the concrete, which are unsafe, unsightly and lead to water loss. Some residents are already doing this in patios as they replace bricks and other hard surfaces with new permeable materials such as Ecosytems grout, which allows water to penetrate into the soil. One architect resident designed a meandering garden, which replaces a concrete area where water runoff is used constructively instead of running down hill in a useless way.

La Luz could better connect with the School of Architecture at the University of New Mexico to foster student awareness of planned area development design and principles of building with the land, not just on it, or in a grid.

La Luz's Architecture Committee has worked to preserve the integrity of the development's original design while adapting to changes in building materials and technologies. La Luz is included on the New Mexico Register of Cultural Properties, the official state list of historic properties worthy of preservation. Home ownership of a cultural property brings potential tax benefits as well as architectural restrictions. Homeowners are eligible to apply for state income tax credits for certain kinds of repairs, updates and modifications, both interior and exterior.[55] For example, a window replacement or a new roof may qualify for tax credits, but a homeowner must apply for approval before renovations in order to receive tax credits. In recent years, residents have been investigating the benefits and drawbacks of seeking inclusion on the National Register of Historic Landmarks. A long-term goal is creating a charitable La Luz Foundation (501C3) to educate all citizens, including developers, about its architectural history, thoughtful use of land, and the values of humane design in planned area development.

The entire community agrees on the critical need for a community center. Originally, Ray Graham's sales office had been designed to serve as a community center, but it remains under private ownership. In the past, two La Luz residents rented it as a shared space for their architectural office and projects. Currently, committees and other groups like La Luz at Home meet in the homes of residents. The Board conducts its meetings in a room provided by the Bosque School. A community center would not only allow a place for meetings, but could also be a place to develop an architecture library, to host lectures, and to conduct small tours of La Luz. A community center could be a place to preserve La Luz's cultural and historical legacy and to educate others about designing planned communities.

As we move into the future we cite the Urban Land Institute, a research source that believes smart cities must maintain a human element in order to properly serve their population. La Luz can serve as an example to others of how to build with, and preserve, the natural landscape and create a park-like setting that enhances the health and psychological well being of its residents. We are indeed fortunate to be part of this far-reaching experiment in design in an artistic and aesthetically pleasing planned area development. As we have maintained in this book, architecture that features good design and incorporates natural and open spaces, as well as shared governance, creates a holistic and healthy environment that nurtures residents and contributes to their happiness and longevity.

We close with the following poetic image that captures the peace of living in La Luz. All houses face the east and mountains, with spectacular views not only for honoring daytime sunrises but also the many and varied moonrises over the Sandia mountain. The glow from a full moon often lights up our evening and night worlds, contributing its ethereal presence to our lives, as captured in this poem by La Luz resident, Marilyn O'Leary:

Lightning Cloud

As twilight came, the cloud was tinged
with pink, big as a small town, high and clear
but soft patterned, scalloped, like a gown.
Then night appeared. The cloud still kept its shape
but lightning, behind or in the cloud itself
pierced the pattern, now and now and now
and from the mountain low the moon arose
at first just lunar light as if contained
by cloud, but then the entire orb appeared,
so full I wondered if the lightning still
would show. Its power was not faded
by the moon. The sky shone with a lightning moon
that made the awesome beautiful, and softened any
fear of Nature's mystery much bigger than
the human scape.
When it was time for bed,
moon bright sky, lightning gone,
I tried to re-picture the awesome play of light
that no camera could save. I wondered
what I really saw, or if
mind's imagining created it, but no—
I trust my sometimes untrustworthy eyes
that let me see the wonder of the skies.

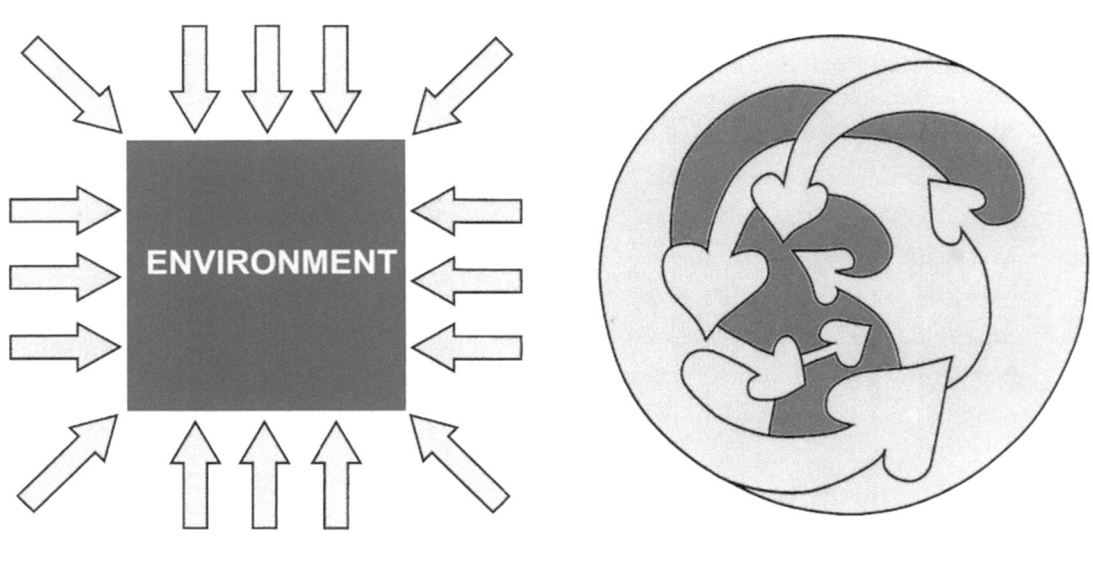

Graphics
(Courtesy of Meredith Taylor)

Afterword

Ecosophy, A New Philosophy Applied to La Luz

The totality of La Luz—its architecture, design, land use, and community spirit—can be understood in terms of a new philosophy—ecosophy. The term originated with Norwegian philosopher, Arne Naess, but it points to an idea not as new as we think. Ecosophy is a philosophy of ecologically responsive design, acknowledging that we live in a complex world of interdependencies, networks, and systems and that we, as humans are a part of, not apart from, those holistic systems. We have a kinship with all things on the planet.

Another example of an architecture embodying a philosophy is Idealism, reflected in the classical architecture of Washington, DC. This formal architecture creates a relationship in which the viewer feels small in relation to larger-than-life statues and buildings, creating a sense of awe for venerable institutions. In contrast, ecosophy acknowledges a reciprocal relationship between humans and their environment. Sensible and beautiful design links humans to nature and to one another. This is the intent of La Luz. This reciprocal relationship with nature, its eco-design, results in a healthy environment for its residents. In this way La Luz as a model can be instructive for urban development.

The study of human factors is important, too, for urban development and its philosophic basis for design. Eco-design not only has been the basis for the design of La Luz in siting, building, solar orientation, lighting, views, and the recycling of materials (in this case, the earth), but also for its sustainability of habitats and the "green" around the architecture and people within this matrix. In such design no object is viewed in isolation, but is seen in its larger context of buildings and people in a reciprocal and loving relation with the land. Eco-design is an intelligent and beautiful system for living, based on a philosophy that has the potential to help us become caretakers and stewards of the planet.

In a way, the architecture has become an educator, providing those who live in such a planned community with a "learning environment" that shows people can not only be stewards of the land, but learners too. La Luz might be viewed as a prototype experiment that the rest of the design world could study in planning the future of cities, especially here in the Southwest with its limited supply of water and other resources. While it is a challenge to keep this philosophy of eco-design alive and growing, in many ways our Native American neighbors have been doing just that for thousands of years. We have a chance to add to that cultural and sustainable record.

Notes

1. M.C. Spires, "City of Adobe and Light: La Luz," *American Home* 73, 50.
2. Ray Graham, interview with authors, January 2018.
3. Didier Raven, interview with authors, May 2018.
4. Raven, interview with authors, May 2018.
5. Antoine Predock, interview with authors, April 2018.
6. Christopher Curtis Mead, *Roadcut: The Architecture of Antoine Predock* (Albuquerque: University of New Mexico Press, 2016), 10, 27.
7. Chris Wilson, *Facing Southwest: The Life & Houses of John Gaw Meem* (New York: W.W. Norton & Co., 2005).
8. James Moore, "Foreword," in *Southwestern Ornamentation and Design, The Architecture of John Gaw Meem* (Santa Fe: Sunstone Press, 1989), 7.
9. Antoine Predock quoted in Collins and Robbins, *Antoine Predock Architect* (New York: Rizzoli, 2000), 13-14.
10. Antoine Predock, interview with authors, April 2018.
11. Kathryn Kaminsky, interview with Cynthia Lewiecki-Wilson, January 2018.
12. Rina Swentzell, "The Relationship Between People and Their Natural and Built Environments," Mass, Journal of the School of Architecture and Planning, University of New Mexico, Albuquerque (Spring 1992), 3-4.
13. Antoine Predock, interview with authors, April 2018.
14. Antoine Predock, interview with authors, April 2018.
15. Quoted in Christopher Curtis Mead, *Roadcut: The Architecture of Antoine Predock* (Albuquerque: University of New Mexico Press, 2016), 27.
16. Antoine Predock, interview with authors, April 2018.
17. Antoine Predock, interview with authors, April 2018.
18. Didier Raven, interview with authors, May 2018.
19. Antoine Predock, interview with authors, April 2018.
20. Antoine Predock, interview with authors, April 2018.
21. Adobe in Action, www.youtube.com/channel/UCumFSAztffkyxQJrZijEOdw, July 29, 2018.
22. Didier Raven, interview with authors, May 2018.
23. Didier Raven, interview with authors, May 2018.
24. Antoine Predock, interview with authors, April 2018.
25. Ray Graham, interview with authors, April 2018.
26. Ray Graham, interview with authors, April 2018.
27. In "La Luz, New Mexico," *Architectural Forum*, 131, 66.
28. Antoine Predock, interview with authors, April 2018.
29. Pueblo Bonito, the largest of the Chaco Canyon great houses, was initially constructed about 850 A.D. Additions in succeeding years show that ancient architecture was connected with the astronomical sky. As J. McKim Malville explains, "Sometime after 1085 C.E., the major axis of the great house was carefully realigned to the north-south meridian" (56) enabling clear views of the winter solstice sunrise (58).

30. Didier Raven, interview with authors, May 2018.
31. Antoine Predock, interview with authors, April 2018.
32. Beth Baurick, interview with Cynthia Lewiecki-Wilson, January 2018.
33. Jonathan Abdalla, interview with Cynthia Lewiecki-Wilson, May 2018.
34. Lynn Perls, interview with Cynthia Lewiecki-Wilson, June 2018.
35. Black Institute for Environmental Studies, Bosque School, www.bosqueschool.org/Black, July 29, 2018.
36. La Luz Open Space Pledge Campaign Update 4/17/04. La Luz Landowners Association.
37. La Luz Articles of Incorporation, New Mexico State Corporation Commission, February 11, 1969.
38. Marianne Barlow, interview with Cynthia Lewiecki-Wilson, July 2018.
39. Jimmie Lueder, interview with Anne Taylor, May 2018.
40. Robert Peters, interview with the authors, May 2018.
41. Robert Peters, interview with the authors, May 2018.
42. Hank Botts, interview with Cynthia Lewiecki-Wilson, June 2018.
43. Lynn Perls, interview with Cynthia Lewiecki-Wilson, June 2018.
44. Laura Campbell, interview with Cynthia Lewiecki-Wilson, June 2018.
45. Patrick Gallagher, interview with Cynthia Lewiecki-Wilson, June 2018.
46. Sandy Masson and Helen Marsee, interview with Cynthia Lewiecki-Wilson, March 2018. Other more common birds they have observed include doves, scrub jays, wrens, several kinds of sparrows and finches, American robins and crows, ravens, and bluebirds.
47. Betsy King, interview with Anne Taylor, May 2018.
48. Jimmie Lueder, interview with Anne Taylor, May 2018.
49. Beth Baurick, interview with Cynthia Lewiecki-Wilson, June 2018.
50. "Alf Simon, landscape architect, estimated that we are losing approximately 900,000 gallons of water from our roofs per year," as reported in "Innovation Sub-Committee Year End Report," La Luz Landowners Association, December 2004.
51. National Resources Conservation Service, USDA, https://www.nrcs.usda.gov/wps/portal/nrcs/main/national/about/, July 29, 2018.
52. Baer invented the "zonohedron," a complex geometric structure that was later confirmed to exist in nature. See www.zomeworks.com.
53. Steve Baer, interview with the authors, June 2018.
54. Collins and Robbins, *Antoine Predock Architect* (New York: Rizzoli, 2006), 13.
55. New Mexico Historic Preservation Division, nmhistoricpreservation.org, July 29, 2018.

Bibliography

Abdalla, Jonathan. Interview with Cynthia Lewiecki-Wilson, May 2018.

Adobe in Action, July 29, 2018, www.youtube.com/channel/UCumFSAztffkyxQJrZijEOdw.

"AIA Honor Awards: Antoine Predock Gold Medalist." Architectural Record, June 2006.

Baer, Steve. Interview with authors. June 2018.

Baker, Geoffrey. Antoine Predock: Architectural Monographs No. 49. Academy Editions. UK: John Wiley & Sons, 1997.

Barlow, Marianne. Interview with Cynthia Lewiecki-Wilson, July 2018.

Baurick, Beth. Interviews with Cynthia Lewiecki-Wilson, January and June 2018.

Black Institute for Environmental Studies, Bosque School, July 29, 2018, http://www.bosqueschool.org/Black.

Bosque School. July 29, 2019. http://www.bosqueschool.org.

Botts, Hank. Interview with Cynthia Lewiecki-Wilson, June 2018.

Campbell, Laura. Interview with Cynthia Lewiecki-Wilson, June 2018.

Collins, Brad, ed. Antoine Predock: Architect. New York: Rizzoli International Publications, Inc., 2006.

———. Antoine Predock, Houses, New York: Rizzoli International Publications, 2000.

Collins, Brad and Juliette Robbins, eds. Antoine Predock, Architect. New York: Rizzoli International Publications, Inc., 1994.

Collins, Brad and Elizabeth Zimmerman, eds. Antoine Predock, Architect. New York: Rizzoli International Publications, 1998.

Collins, Brad and Elizabeth Zimmerman, eds. Architectural Journeys: Antoine Predock. New York: Rizzoli International Publications, 1995.

Dethier, Jean. Adobe Architecture: An Old Idea, A New Future. Down to Earth. Based on the Exhibit Des Architectures des Terre, Centre Georges Pompidou, Paris, 1981. First published in the USA, Facts on File, 1983.

Dixon, John Morris. "Regionalism in the Southwest." Progressive Architecture (March 1974): 68-69.

Gallagher, Patrick. Interview with Cynthia Lewiecki-Wilson, June 2018.

Graham, Ray A., III. Interviews with authors. January and April 2018.

Innovation Sub-Committee Year End Report, La Luz Landowners Association, December 2004.

Kaminsky, Kathryn. Interview with Cynthia Lewiecki-Wilson, January 2018.

King, Betsy. Interview with Anne Taylor, May 2018.

Kroloff, Reed. "La Luz: Quarter-Century Accord with Nature." Phoenix Home and Garden (February 1994): 41-43.

La Luz Articles of Incorporation. New Mexico State Corporation Commission. February 11, 1969.

"La Luz, New Mexico." Architectural Forum 131: 66-71.

La Luz Open Space Pledge Campaign Update 4/17/04. La Luz Landowners Association.

"La Luz—Stadt in Der Sonne (La Luz—City in the Sun)." Schoner Wohnen Magazine, March 1984.

Lueder, Jimmie. Interview with Anne Taylor, May 2018.

Malville, J. McKim. A Guide to Prehistoric Astronomy in the Southwest. Boulder: Johnson Books, 2008.

Mann, Martha. "Happy 25th La Luz Vision Shines at Gala." Albuquerque Journal (Sunday September 12, 1993): D1, 3.

Marsee, Helen. Interview with Cynthia Lewiecki-Wilson, March 2018.

Masson, Sandy. Interview with Cynthia Lewiecki-Wilson, March 2018.

McHarg, Ian. Design with Nature (1969). Natural History Press, 1971.

Mead, Christopher Curtis. Drawing into Architecture. Albuquerque: The University of New Mexico Press, 2016.

———. Roadcut: The Architecture of Antoine Predock. Albuquerque: The University of New Mexico Press, 2011.

Moore, James. "Foreword." In Southwestern Ornamentation and Design, The Architecture of John Gaw Meem, by Anne Taylor, 10. Santa Fe: Sunstone Press, 1989.

National Resources Conservation Service, USDA, July 29, 2018, https://www.nrcs.usda.gov/wps/portal/nrcs/main/national/about/

New Mexico Historic Preservation Division, July 29, 2018, nmhistoricpreservation.org.

O'Leary, Marilyn. "Lightning Cloud." Unpublished peim, 2018.

Pearson, Clifford A. "AIA Honor Awards. Antoine Predock: Gold Medalist." Architectural Record, June, 2006: 223-224 + illustrations.

Perls, Lynn. Interview with Cynthia Lewiecki-Wilson. June 2018.

Peters, Robert. Interview with the authors. May 2018.

Predock, Antoine. Interview with authors. April 2018.

Price, V.B. "La Luz: The Light on the Mesa." Su Casa (Autumn, 2001): 34-39, ff.

———. "Shaping the New Southwest." Su Casa (Summer, 2006): 88-96.

Pulkka, Wesley. "Antoine Predock: Architecture and Time." Trend + Art + Design (Fall 2009-Winter 2010): 56-63.

Raven, Didier. Interview with authors. May 2018.

Spires, M.C. "City of Adobe and Light: La Luz." American Home 73: 50-55.

Swentzell, Rina. "The Relationship Between People and Their Natural and Built Environments." Mass, Journal of the School of Architecture and Planning, The University of New Mexico, Albuquerque (Spring 1992): 3-5.

Taylor, Anne. Linking Architecture and Education: Sustainable Design of Learning Environments. Albuquerque: The University of New Mexico Press, 2009.

———. Southwest Ornamentation and Design, The Architecture of John Gaw Meem. Santa Fe, New Mexico: Sunstone Press, 2017.

Thompson, Elisabeth Kendall, ed. "La Luz Plaza Fountain." In Apartments, Townhouses, & Condominiums, second edition. McGraw-Hill,1975.

Tudge, Colin. The Secret Life of Trees. UK: Penguin, 2006.

Urban Land Institute. July 29, 2018. https://uli.org/

Wilson, Chris. Facing Southwest: The Life & Houses of John Gaw Meem. W. W. Norton & Co., 2005.

Zomeworks. July 29, 2018. www.zomeworks.com

Index

Albuquerque, New Mexico, 13, 15, 16, 36, 46, 61, 64; West Side Coalition, 46, 64.
Ancestral puebloans, 13, 14, 16. *See also* Chaco Canyon. See also Pueblos.
Antoine Predock Center for Design and Research, 10. *See also* The University of New Mexico.
Architecture, 13, 17; building materials, 13, 14, 16-17, 18, 22, 26; eco-design, 73; styles, 15, 17-18, 19, 68.
Architectural Forum, 38.

Baer, Holly, 68.
Baer, Steve, 16, 17, 68.
Barker, Hildreth, 31; Barker Bol Architects, 31.
Borkovets, Richard, 47, 61.
Bosque del Apache, 51.
Bosque School, 42, 44, 67, 69, 70; BEMP program, 42; Black Institute for Environmental Studies, 42.
Brand, Stewart, 17.
Browne, Graham, 45.

Chaco Canyon 13, 19, 38; Pueblo Bonito, 14, 38.
Chavez, John I., 31.
Columbia University, 17.
Coronado (Francisco Vásquez de Coronado), 16.
Corrales, New Mexico, 16-17, 68.

Dahlquist, Gunnar, 15, 27, 30.

Ecosophy, 73.
Ecosystem, 69.
El Paso, Texas, 14.

Frampton, Kenneth, 24.

Ghost Ranch, New Mexico, 56.
Graham, Barbara, 16.
Graham, Ray A., III, 15, 16, 17, 41, 42, 45, 61, 70.

Handbook for La Luz Landowners, 47.

Isleta Pueblo, 16-17. See also Pueblos.

Kahn, Louis, 17.
Kinney Brick Company, 30.

"landscrapers," 10.
La Luz Landowners Association, 23, 45, 46-50; LLLA Board of Directors, 63, 70.
La Luz Voice, 48.
Las Cruces, New Mexico, 42.
La Villa Real del la Santa Fe de San Francisco de Asis, 14. *See also* New Mexico history.

McHarg, Ian, 20.
Mead, Christopher Curtis, 17.
Meem, John Gaw, 18-19.
Moore, James, 19.

Naess, Arne, 73.
National Register of Historic Landmarks, 70.
New Mexico, 13, 17, 38; history 14, 15, 16. See also architecture. *See also* pueblos.
New Mexico Register of Historic Landmarks, 70.

Ovenwest Corporation, 15, 23, 45.
Ohkay Owingeh Pueblo, 42. *See also* Pueblos.

O'Leary, Marilyn, 71.

Predock, Antoine (Tony), 15, 17, 19, 23-24, 26, 29, 38, 41, 42; awards 10.
Pueblos (New Mexico), 13-14, 16-17, 20, 21, 42, 48. *See also* Ancient puebloans.

Raven, Didier, 16-17, 24, 27, 41, 45, 68.
Rio Grande (river), 14, 23, 42.
Rio Grande Nature Center, 69.

Sanchez, Ernie, 26.
Sandia Mountains, 20, 24, 34, 35, 71.
San Felipe Pueblo, 30. *See also* Pueblos.
Santa Fe, New Mexico, 15
Schlegel, Don, 17, 27,
Skidmore, Owings, and Merrill, 48.
Socorro, New Mexico, 51.
Swentzell, Rina, 21.

Taos, New Mexico, 15; Taos Pueblo, 20, 48. *See also* Pueblos.
Taylor, Joel, 17.

United States Department of Agriculture Natural Resources Conservation Service, 67.
The University of New Mexico, 18, 42, 70; UNM School of Architecture and Planning 17, 70; UNM Law School building, 17, 20.
Urban Institute, 41.
Urban Land Institute, 70.

Wright, Frank Lloyd, 17.
Wright, George, 20.

Zome, 17, 68; Zonohedra 68.

About the Authors

The authors, Anne Taylor and Cynthia Lewiecki-Wilson
(Courtesy of Jennifer Fenstermacher)

Anne Taylor, PhD. University of New Mexico Regents and Association of College Schools of Architecture (ACSA) Distinguished Professor of Architecture and Planning, Emerita, attended Wells College and Arizona State University where her research area was human factors and the effects of the physical environment on learning and behavior of four year olds and English language learners. Through the years she has worked with architects to program and design newer schools and has developed an integrated design education curriculum installed in many schools internationally for use with teachers and children pre-kindergarten through high school: "Architecture and Children Core Curriculum" (now translated into Japanese, Spanish, and Turkish), "Architecture and Children Southwest, Architecture and Children Northwest," and with Duvall, "Integrated Design Education for Second Grade." Taylor is the author many articles and six books among which are Linking Architecture and Education: Sustainable Design of Learning Environments and Southwestern Ornamentation and Design, The Architecture of John Gaw Meem that can be used by teachers, architects and others to explore with their students symbolic architecture and design thinking in a unique part of the United States.

Cynthia Lewiecki-Wilson, PhD, now retired, was a Professor of English at Miami University of Ohio, where she taught writing and rhetoric, literature, women's studies, and disability studies courses. She directed a writing program as well as the Graduate Programs in English and founded the disability studies minor at Miami. She was one of the founding members of disability studies in rhetoric, which has since grown into an international interdisciplinary field. She has served on the editorial boards of national and international journals, and has published over thirty scholarly articles and chapters. She is the author or co-editor of five books: Writing against the Family: Gender in Lawrence and Joyce (Southern Illinois University Press), From Community to College: Reading and Writing across Diverse Contexts (St. Martin's Press) with Jeff Sommers, Embodied Rhetorics: Disability in Language and Culture (Southern Illinois University Press) with James C. Wilson, Disability and the Teaching of Writing (Bedford/St. Martin's) with Brenda Jo Brueggemann, and Disability and Mothering: Liminal Spaces of Embodied Knowledge (Syracuse University Press) with Jen Cellio.